ROUTLEDGE LIBRA
20TH CENTURY

T0227498

Volume 13

WORLD CHAOS

WORLD CHAOS
The Responsibility of Science

WILLIAM McDOUGALL

LONDON AND NEW YORK

First published in 1931

This edition first published in 2014
by Routledge
2 Park Square, Milton Park, Abingdon, Oxon, OX14 4RN

and by Routledge
711 Third Avenue, New York, NY 10017

Routledge is an imprint of the Taylor & Francis Group, an informa business

First issued in paperback 2016

British Library Cataloguing in Publication Data
A catalogue record for this book is available from the British Library

ISBN13: 978-1-138-01361-2 (hbk)
ISBN13: 978-1-138-98736-4 (pbk)

Publisher's Note
The publisher has gone to great lengths to ensure the quality of this book but points out that some imperfections from the original may be apparent.

Disclaimer
The publisher has made every effort to trace copyright holders and would welcome correspondence from those they have been unable to trace.

WORLD CHAOS

THE RESPONSIBILITY
OF SCIENCE

BY

WILLIAM McDOUGALL

F.R.S., author of *An Introduction to Social Psychology*, etc.

LONDON

KEGAN PAUL, TRENCH, TRUBNER & CO., LTD.

1931

PRINTED IN GREAT BRITAIN BY
STEPHEN AUSTIN & SONS, LTD., HERTFORD

PREFACE

Few will deny that the world is in a state to provoke the laughter of the gods. Our Western Civilization, of which we have been so proud, is responsible for the threatening chaos. With it lies the world's only hope of achieving some tolerable degree of order, harmony and prosperity within any period in which we and our children can feel any vivid interest.

The distinctive feature of our civilization is our Science. The thesis of this little book is twofold : first, that physical science has been the principal agent in bringing about the very rapid changes in our social, economic and political conditions which are the source of our present troubles ; secondly, that in the development of the neglected social sciences lies our only hope of remedy for those troubles.

The physical sciences, while conferring many benefits, have produced a complexity of our civilization which far outruns our present understanding and power of control. We may hope to attain the much needed understanding in time to avert the ultimate, and probably the rapid, decline of our civilization, only if we take in hand, energetically and with resources at least equal to those hitherto devoted to the physical sciences, systematic research in the sciences of man and society. Time presses, and I have written with ruthless disregard of the feelings of my scientific colleagues. To them my apologies are due.

The main substance of the book was delivered as the Ludwig Mond lecture before the University of Manchester in May of this year. I have added a number of passages illustrating and enforcing the very concisely stated propositions of the lecture. These supplementary passages are insetted to distinguish them from the substance of the lecture, which may be read in its original form by those who prefer a diet without roughage.

W. McD.

Grasmere.
August, 1931.

WORLD CHAOS

THE RESPONSIBILITY OF SCIENCE

THE civilization of any people reflects the state of its knowledge, and is in a large measure determined by that knowledge. If, then, we have reason to be profoundly dissatisfied with the state of our civilization, we shall do well to consider whether there is not some radical defect in our knowledge, more especially in the systematically organized part of our knowledge which we call Science.

It would not be true to say that our Western Civilization is founded upon Science; for its deeper foundations are the curiously blended traditions of Christianity and of the classical world of Greece and Rome. But modern Science, the Science that dates from the Copernican revolution, the Science founded by Copernicus, Galileo, Descartes and Newton has profoundly modified it, has given it a highly peculiar quality, a quality that tends to overshadow, to obscure and even to destroy those ancient foundations. For the most distinctive quality of our present civilization is that it undergoes perpetual and rapid change and that its ideal is progress rather than stability; and this quality is the gift of Science.

If all seemed well with us, if peace and material prosperity seemed assured to all nations, the present state of our civilization would nevertheless justify some misgivings ; we might well ask, Can changes, so great, so rapid, so continuing, be compatible with stability ? We are building up an immense and very complicated superstructure. May it not be top-heavy ? May not the foundations be crushed and crumbled, until the whole structure erected on them must collapse ?

It was, no doubt, a sense of this increasing top-heaviness of our civilization that prompted an English bishop recently to suggest that Science should take a holiday, that for half a century, at least, all scientific research should be suspended while we consolidate our gains and make sure of our foundations. The suggestion is an impracticable one. We cannot stop and stand still even if we would ; we are committed to further progress or, at least, to further change. The knowledge and power which Science has given us must inevitably carry us on with a frightful momentum. We are like men in an airplane crossing a great ocean ; and the driving power is Science. To cut off the engine in mid-ocean could result only in disaster. Though we know not where we are, nor whither we are going, we must keep on, hoping for the best and applying the discoveries of Science to the best of our ability and our wisdom.

A complete cessation of all scientific research would, no doubt, result in a slower rate of change than is likely under the influence of continued and increasing research. But the process of change has a momentum of its own which must carry it on, perhaps at an accelerating pace, even without the influence of further research. The further diffusion of the knowledge we already have, the wider technical application of that knowledge, the many social and political changes already in progress are consequences of the application of scientific knowledge and of the undermining of old traditions and beliefs by that knowledge ; all these consequences will inevitably continue to unfold themselves, producing new complications, new problems, new dangers, as well as new benefits. Consider the knowledge of methods of birth-control. This knowledge, now let loose upon the world irrevocably, may within a few generations result in changes of the composition of the world's populations far more violent and momentous than any recorded by history, than any resulting from the most devastating plagues, the most destructive wars, floods or other catastrophies. And the changes of population that may well result from this new knowledge are likely to be far more profound and lasting, and only too likely to be far more disastrous, than any previous changes of population ; because, while in the main those former changes were non-selective, affecting all parts and degrees of each population equally, or, if selective, then, in the main, favourably selective, the changes due to the new factor are likely to be highly selective and so far as can be foreseen, in the absence of public regulation of some sort, selective in a highly unfavourable sense.

Consider the influence of the means of easy transportation of human beings in large numbers over great distances. This factor has been in operation for only a short period, yet already it has effected changes in the distribution of races and peoples in comparison with which all earlier migrations vanish into insignificance.

Consider the effects of " radio " and of the " cinema ", now just beginning to make themselves felt in all parts of the world. Who can foresee the extent and nature of their influence ? All we can certainly foresee is that they must be very great and profound.

Consider the effects of the spread of large-scale machine-production among the populations of Russia, India and China. Already the inception of the process in Russia is producing profound uneasiness in the nations already industrialized ; just because the world-effects must be so vast and are so unpredictable.

The proposals of Mr. Gandhi and of other enthusiasts for the simple life come too late. It is impossible to forswear the methods of machine civilization and to return to the simple economy of hand-production and of agriculture with primitive tools, if only because, by the aid of machines driven by steam and electric power, we have multiplied immensely the numbers of mankind. Mr. Ghandi's ideals, if they could be put into practice throughout India, would entail the reduction of the population by some two hundred millions. The lives of vast numbers of mankind are increasingly dependent on the unremitting diligence of the highly trained experts who maintain in working order, by the aid of scientific knowledge, the vast systems of power-supply, of transport and of communication, and only in less degree upon those who maintain sanitary services, manufactures, and the protection of animal and plant life against pests and plagues which under modern conditions would, if uncontrolled, threaten the destruction of our principal food-supplies. The spread of infectious diseases of animals and plants, as well as of human beings, from one continent to another has been one of the most troublesome consequences of the modern facilities of transportation. The transported disease-producing organisms find a population that, having little or no power of resistance to its ravages, falls an easy prey. This has been very troublesome in America, where the trees of some one species have in a number of instances been well-nigh exterminated over vast areas in the course of a few years.

I have said that our civilization is top-heavy : and I have suggested that this is a defect and a danger which we can do little or nothing to remedy. The foundations were laid long ago, the superstructure exists as a going concern and we must keep it going. We cannot refuse to use the immense resources which Science has put at our dispoal.

A second characteristic of our civilization is no less a defect and a danger than its top-heaviness ; namely, it is lop-sided. Now a top-heavy structure may stand firmly erect if it is symmetrical and well balanced ; but, if it is both top-heavy and lop-sided, disaster threatens. And not only is our civilization both top-heavy and lop-sided ; it grows more top-heavy and lop-sided with every year that passes ; consequently its stability becomes more doubtful, and the danger grows more threatening. I have said that we can do little or nothing to make our structure less top-heavy; fortunately it is possible to make it less lop-sided, and thus to give it greater stability and, perhaps, at the same time to attach it more firmly to its foundations.

This second defect and the remedy for it are the topics of this brief discussion.

Our Civilization is Lop-sided

The top-heaviness of our civilization is due to the rapid development of Science ; its lop-sidedness is due to the lop-sidedness of our Science. Our civilization reflects the state of our knowledge ; and especially it reflects it faithfully in respect of the lop-sided state of our Science.

That our Science is very lop-sided is indisputable, a matter of common agreement. Since the time of Galileo, Physical Science, by which I mean the sciences of the inorganic or physical realm, has advanced at a constantly accelerating pace. The Sciences of life have lagged far behind. Until the middle of the nineteenth century they were rudimentary, concerned merely with description and classification. The work of Charles Darwin gave them a fillip. For a time it seemed as though they were about to progress rapidly ; it was even confidently prophesied that, as the nineteenth century had been the century of physical science, so the twentieth century was to be the century of the biological sciences, was destined to go down in history as the great age of biological discovery. We are now near the end of the first third of the century, and it seems very unlikely that this prophecy will be realized. Physical science still accelerates its progress. The biological sciences limp painfully behind. If this is true of biology proper, still more is it true of the sciences of man and of society ; we talk of psychology, of economics and of political science, of jurisprudence, of sociology and of many other supposed sciences ; but the simple truth is that all these fine names simply mark great gaps in our knowledge, or rather fields of possible sciences that as yet have hardly begun to take shape and being. The names stand for aspirations rather than achievements ; they define a programme, they vaguely indicate regions of a vast wilderness hardly yet explored, and certainly not mapped, regions in which chaos still reigns, yet *regions which must be reduced to order if our civilization is to endure.*

The grounds of this lop-sided state of our Science it is easy to indicate. First, the physical sciences are the easier : as M. Bergson has so forcibly pointed out, their problems are such as our minds are adapted to deal with. The mental capacities of the human species have evolved in the course of a struggle for existence, in which struggle the first condition of survival has been effective dealing with the material environment ; and in spite of all the fine things said about the scientist's pure love of truth, the truth seems to be that the main spur to the development of physical science was man's practical need of understanding and control of his physical environment.

The success of the physical sciences in bringing such understanding and control has been and still is the second great ground of the predominance of those sciences in our field of knowledge. Physical science has brought such great and obvious benefits to mankind that it commands the respect, the admiration and the ungrudging support of all the civilized portion of the human race. I need not dwell upon these benefits, the multitude of conveniences, comforts and luxuries. Look about you, consider the contents of your breakfast table, or your parlour with its electric light and radio-outfit ; or penetrate to remote jungles of Malaysia and see the naked savage using lucifer matches from Japan, cotton cloth from Manchester, pottery from Staffordshire, a rifle from Hartford, Conn., and, perhaps, spectacles from Birmingham.

The prestige accruing to physical science from the practical benefits it has conferred is the greater because those benefits are in the main obvious and unquestionable ; and the fact that they have resulted directly from scientific research is easily grasped by the common man. The steam-engine, the motor-car, the rifle, the radio, the telegraph and telephone, and a thousand other mechanisms in common use perpetually remind the common man of his debt to physical science. But, when he eats cheap and good and varied food gathered from remote parts of the world, nothing reminds him that biological research has contributed greatly to make this possible. Still less does he realize that, if his expectation of life and health is prolonged by some twenty years beyond that of his grandfather, this is the consequence of biological and medical research. And when he reaps the benefits of living in a well-ordered peaceful community in which his rights are protected against the rapacity of rulers and the greed and lust of his fellows, in which vastly complicated systems of finance, of education and of parliamentary representation are maintained in smoothly working order, he is a very exceptional person if he at all realizes that all this has been rendered possible only by a vast amount of thought and discussion and research in the field of the sciences of man and of society.

A third ground of the backwardness of the biological and human sciences has been the opposition of the Churches. The Church quickly learnt to adapt its doctrines to the Copernican revolution and to the teaching of Newton ; and it welcomes with open arms and loud acclaim the Einstinian revolution.[1] But, though it has permitted the study of some of man's cultural achievements, such as language and literature, it has frowned upon the study of his beliefs, his superstitions and his religions, his magic and ancient customs ; and especially it has opposed all more direct approach to the problems of human nature, as well as all biological studies (such as the dissection of the human body and the comparative study of men and animals) which tend to reveal man as a part of Nature. And, let me remind you, our Universities have been,

[1] I see the Einstein principles are invoked just now to render more easily credible the story of the Ascension.

and still are, largely controlled and shaped by the Churches.

A fourth ground of the rudimentary state of the biological and human sciences is perhaps the most important at the present time.

The long start and the relatively advanced state of the physical sciences, together with the great prestige accruing to them from their brilliant successes, have worked to the great detriment of the sciences of life.

And this in two ways. First, the methods and principles that have proved so successful in the physical sciences have been accepted by a great majority of men of science as alone valid for all sciences ; and the account of the world rendered by Physics has been accepted by most men of science and by many philosophers, from Spinoza onward, as literally true. The work of modern philosophers has consisted very largely in fruitless efforts to reconcile with the mechanistic principles of physical science and with the account of the world rendered by it some belief in human values and in human efforts to conserve and augment these values. The biologists, accepting physical science as the model of all true science, have for the most part been resolutely blind to the abundant evidences of the causal efficacy of mental activity in the organic world. This general acceptance of the mechanistic account of the world has hampered and perverted all biological and, especially, all psychological research. And, like a dark cloud, it has enveloped and overwhelmed the popular mind ; has impressed and oppressed it.

The depth of this impression and oppression has been vividly illustrated by the world-wide applause which has greeted recent announcements by physicists that physical science has been in error. When physicists, like Jeans and Eddington, have recanted the major errors of the scientific dogma, the world at large, instead of turning and rending them with reproaches, instead of crying " Then why have you physicists so long oppressed us with your nightmare dogmas ? ", still grovels before physical science and regards these recantations of error as announcements of great scientific discoveries. So great is the prestige of physical science in the popular mind !

The vast prestige enjoyed by the physical sciences has not quite the same grounds in the scientific world and in the popular mind. No doubt in both cases the brilliant successes of physical science in giving us control over natural processes and in adding to the conveniences and comforts of life has been the main factor. But the scientific mind has been fascinated by the enterprise of explaining all the processes of the universe, from the infinitely large scale of sidereal events to the ultra-microscopic atomic dance, by one system of simple and, in a sense, intelligible principles, the Newtonian principles of mass and momentum. Dazzled by the prospect, inspired with enthusiasm for the forwarding of this great enterprise, the majority of scientists have, until very recently, been strangely blind to the great wealth of phenomena that resist all such attempts and show no signs of yielding to them. It is not only that, as Dr. J. S. Haldane has recently repeated, " physical science cannot express or describe biological phenomena, so that its claim to represent objective reality cannot be admitted " ; it is also that the mechanistic scheme of explanation breaks down within the sphere of inorganic phenomena, as many physicists are now at last beginning to see. But, in spite of this new light among the physicists, Dr. Haldane's characterization of our age remains true :

" We are still living in an age which I think our successors will some day look back upon with curiosity and wonder as an age characterized especially by physical realism—an age strangely blind in some, but by no means all, respects to what will then appear as outstanding spiritual reality, and concealing this behind scientific abstractions which it had taken for representations of reality and proceeded to bow down before, though they were only its own creations. In this respect I think that our age will be regarded as an idolatrous one, although our idols are of a different kind from those of relatively uncivilized peoples. We can see signs of the passing of this idolatrous age of physical realism or materialism, but how long it will take to pass no one can say." [1]   In this system of idolatry the physical scientists play the part of high priests, imposing their dogmas upon the public in perfect good faith; and the public, including its more educated part, receives their teaching with primitive credulity. It can hardly do otherwise. Apart from all the prestige resulting from its practical

[1] *The Philosophical Basis of Biology*, London 1931.

applications, physical science seizes and holds the imaginations of the public, filling them with awe by its descriptions of vast sidereal spaces and events, by its predictions of final freezings or of final burnings, and with creepy curiosity by its descriptions of atoms magnified to the size of cathedrals. If the majority of physicists have long remained blind to the inadequacies of the mechanistic scheme in the inorganic world, and if the majority of biologists are still capable of believing that the mechanistic scheme is, in principle, adequate to the interpretation of living things, how inevitable that the public at large shall be misled and shall continue to accept, whether resentfully and coweringly, or with whole-hearted adulation of its high priests, the mechanistic world-scheme as substantially correct ! No wonder that the eloquent and dogmatic re-affirmation of the mechanistic creed in the recent book of Professor L. Hogben [1] has been widely acclaimed in the Press ! No wonder that he, an outspoken mechanist, has been appointed to fill the new and highly important chair of social biology in the

[1] *The Nature of Living Matter*, London, 1930.

University of London ! He is appointed, presumably, in the sure hope that by chemical analysis of the blood or other tissues of our legislators he will succeed in solving the economic and political world crisis, or will at least prevent his colleagues, the social scientists, straying from the firm basis of mechanistic dogma on which they have been brought up.

A recent symposium of philosophers at Cambridge afforded amusing and instructive illustrations of what I am saying about the prestige of physics and physicists. In this symposium on determinism, Sir Arthur Eddington took part. He revealed that his view of the universe had been profoundly transformed in the course of the last five years, and he very naturally and, in the main, correctly assumed that the view entertained by the world in general had undergone, or must necessarily undergo, a similar change. One philosopher after another got up and politely assured the distinguished astronomer that what he was now announcing was stale news to them, and that the views that happen to be in fashion among physicists do not really determine the structure of the universe.

But Sir Arthur was right in his assumption that those views determine, if not the structure of the universe, the public beliefs about it. Philosophical criticism has in the main passed over the heads of the public. And many modern philosophers have shown an indecent haste to harmonize their metaphysics with the dogmas of physical science. Thus the Hegelian system, which, up to the time of the Great War, was predominant in our universities, was there so manipulated as to make this system of pure idealism seem to coincide or harmonize with a strictly mechanistic interpretation, not only of stars and atoms, but also of living things, animals, men and nations.

Another important factor in maintaining the preponderance of the physical sciences is the relatively short preparation required for effective research in their fields. A young physicist may hope to make within a few years of graduating from college some crucial observation which will bring him world-wide fame. The biologist, and still more the social scientist, is in very different case. After graduating from college the latter needs many years of assiduous study to prepare him for effective research. He needs a general acquaintance with all the fundamental sciences, and, especially, a good grounding in biology ; he must also make himself a psychologist, a task hardly to be accomplished by the most gifted by less than five or perhaps ten years of steady work ; he needs to read widely in history and in the voluminous literature of the social sciences, a veritable jungle where he can hope for no sure guidance and no fixed landmarks.

To all this may be added, as a not unimportant factor in maintaining the public prestige of the physicists, the practice by them of what in less dignified callings would be described as efficient log-rolling. Divided as they are on some of the most fundamental questions, the physicists stand loyally by one another, and the fact of their wide differences largely escapes the public. Not infrequently one physicist criticises another severely, saying in effect that all his views and utterances are mere nonsense, but loyally adds " Such is my opinion, but, of course, Sir A.— B— or Lord C. (as the case may be) is a very very great scientist, whose shoes I am unworthy to unlatch." In strong contrast is the acrimonious depreciation of both forerunners and contemporaries which is common form among the followers of the human sciences. Newton is still held in highest honour at Cambridge. John Locke, who may fairly be said to deserve equal fame in his different sphere, is without honour in his own university.

Further, a vicious circle has been established. Physical science, having obtained a long lead and a vast prestige, continues to increase her lead and her prestige at the cost of the sciences of life; for, as we know, nothing succeeds like success. The universities provide palatial accommodation and unlimited equipment and a multitude of academic posts for the workers in the physical sciences. In addition, great industrial corporations spend vast sums on physical research. It is officially estimated that in the United States alone research (almost exclusively in the physical sciences) is subsidized by industry to the extent of five hundred million dollars a year, a sum considerably greater, I suppose, than the combined incomes of all British Universities. Lord Rutherford (*The Times*, 9th June, 1931) is quoted as saying that tens of thousands of men are to-day engaged in research directed to improvement of the motor-car ; and it has recently been alleged that, while in America 200 chemists are devoting their energies to the discovery of more deadly poison gasses, this country is spending £200,000 a year in the same noble cause.

Consider now the state of our civilization. It reflects, as I said, the state of our knowledge or science ; it is therefore characterized by its multitudinous applications of the discoveries and inventions made by, or made possible by, the physical sciences, applications which have transformed our world and our relations to one another, political, social and economic, national and international. The total effect is commonly described, not quite accurately, yet not quite unjustly, as an increasing mechanization of all our civilization.

The principal feature of the change wrought by physical science is that we have at our disposal vast stores of physical energy ; which energy we have applied in two ways. First, as a substitute for the muscular energy of men and domestic animals in the production of the fundamental necessities of human life, food, shelter and clothing, with the immediate consequence that the population of the world has multiplied as never before. And it is important to notice that this very rapid multiplication has taken place not only in those peoples who have developed the physical sciences and their applications, but also, in only less degree, among the vastly more numerous peoples who have had no active share in that development. I remind you, merely as an illustrative instance, of the fact that the population of India was multiplied roughly threefold during the nineteenth century.

I have discussed elsewhere[1] these facts and their bearings at some length, showing the danger that the present tendencies, if uncorrected, may well result in a swamping of the white " race " by the swarming multitudes of other races that have not yet proved their capacity to sustain, much less develop for themselves, any high state of civilization. Here I will only point out that we are now actively engaged in promoting in Africa a vast increase of the Negroid population, an increase comparable to that which has already resulted in India. Whether it is desirable that Africa should

[1] *Ethics and Some Modern World Problems*, London, 1924.

contain a Negroid population of many hundred millions is a problem on which I express no opinion. I merely point out that it is a problem that has not yet reached the plane of discussion whether by statesmen, philosophers, or the general public. It seems unlikely to reach that plane until the world shall be faced with the *fait accompli*. Nevertheless the responsibility lies with us of the present generation ; our actions are setting the vast change in process, and we are acting in ignorance and without even attempting to consider the probable consequences for mankind in general.

Secondly, we have produced a multitude of comforts and conveniences which have become so intimately woven into the texture of our civilization that they are now essentials ; the deprivation of them would cause not only much discomfort and suffering but also the breakdown of the whole complex structure of our civilization ; without them we should starve and die by millions in all parts of the world. That is to say, the discoveries made by physical science have greatly increased the numbers of mankind and have added immensely to the complexity, the delicacy, the intimacy, the vital importance of the relations between men and between groups of all kinds.

"Mass production" is the latest great step in the revolution produced by physical discovery. In itself it may have the profoundest consequences in the near future for the life of mankind in general and for the fate of various nations. On this topic hear a great authority. Sir Eric Geddes (*Observer*, 14th June, 1931) writes :—

"Everyone of us who has lived for the thirty years of this century has seen a series of revolutionary changes in the world—in international relations, politics and finance, and in the realm of invention. In the industrial field, from the production point of view, a revolution has taken place which I think ranks second to none, not even to the effects of the great war, in its growing reactions upon the human race. I refer to the development of mass production in its widest sense and in its varying degrees. It may mark and make the rise and fall of nations. Mass production was commencing before the war. Since the war it has developed with great rapidity. . . . Mass production has, with its low price, brought articles of convenience and luxury into lives of untold millions who, last century, could never have dreamed of being able to afford such things. Its power for the well-being of mankind is enormous. It may be that the influence of mass production upon the destinies of Great Britain will be far greater than that of the war and adventures from which the British Empire grew."

Meanwhile "mass production" is producing "technological unemployment" on a vast scale, with all its attendant mass of suffering ; and, hitherto, its effect upon the state of Great Britain has been greatly to worsen rather than better it.

Mass production is also responsible for the over-production, both in agriculture and industry, which is a chief cause of the present fall of prices and the world-wide economic depression. Could there be a more striking demonstration of my thesis that lack of understanding and control of the human and social factors of our civilization lags far behind our material development and renders nugatory and even gravely injurious advances in physical knowledge and control which might be of the greatest benefit to all the world. The same is true of that intimate financial, economic and political interdependence of all parts of the world which is a characteristic novelty of our age. It is much celebrated as a great step towards universal brotherhood ; but actually it is a most serious threat to such harmony, prosperity and stability as the world can show. It was the complexity of international relations and interests, especially those between Germany and Great Britain, that brought upon us the war of 1914. And at the present time the relations between America and Great Britain are assuming an ominous similarity to the pre-war relations of Great Britain and Germany. This is brought out very clearly and forcibly by Mr. Ludwell Denny in his abundantly documented work, *America Conquers Britain*.[1] He writes :—

"Unfortunately most of the peace movements of both America and Britain rely almost exclusively on safeguards other than an understanding and settlement of basic economic conflicts. They ignore or deliberately gloss over those realities. They

[1] N.Y., 1930.

see Utopia in terms of treaties and the kind of naval limitation or truce which passes for disarmament. Certainly no intelligent person can belittle the temporary settlement of political and naval disputes resulting from economic conflict. But neither can he assume that the cause is thereby eliminated." . . . "Treaties are not enough, not even an unconditional arbitration treaty, which is much needed. . . . If ever they refuse to fight, if ever they refuse to believe the propaganda and war lies of governments, if ever they decide that the actual faults of the 'enemy' cannot be corrected on the battlefield, it will be because they understand the nature of the conflict. Armed with knowledge, they may not fight with guns. Those who preach the unthinkableness of war between the United States and Britain, those who pray for silence regarding the present Anglo-American economic conflict, have perceived a half-truth, but a most dangerous half-truth. They see that public opinion is the only hope. But they do not see that uninformed and unintelligent public opinion is the great menace. . . . British and American public opinion in its present state could not be trusted. There is danger of eventual war. There is a fierce struggle for foreign markets, raw materials, financial supremacy. But that struggle in itself is not the gravest danger. The danger is in the people's ignorance. They believe that international conflicts can be settled by armies and navies. They still believe that a war can be won."

Mr. Denny implies that the great source of danger is the ignorance of the man-in-the-street. I differ from him only in holding that the ignorance that is our danger is deeper going, is the ignorance of all of us, of experts and statesmen no less than of the masses, that we need not merely the diffusion of such knowledge as is available but also, and more importantly, a great increase of the knowledge to be diffused.

In all former ages the relations of man to man and group to group (the civilized no less than the savage) were governed by custom and tradition, law being merely the formal recognition of custom and tradition. The family, the clan, the tribe, the kingdom, the feudal system, the parliament, all such institutions expressed and were adequately governed by old traditional loyalties. Under the vast complexity of modern conditions, this old traditional wisdom is utterly inadequate to regulate our relations. We are compelled to try to live by the light of Science ; and alas ! we have no Science to guide us. *The physical science which has produced this new complexity can give us no guidance whatsoever in our difficult task of coping with it.*

Consider our plight. The family, the most deeply rooted of all our traditional institutions, the foundation of all the rest, decays ; and we are threatened with general deterioration of the peoples that have created our Western Civilization, if not their actual extinction or substitution by other races.

Here, as in so many other matters, America leads the way and points the path along which the rest of our Western Civilization seems destined to wander. It is in the United States that the applications of physical science have been made on the greatest scale, and it is there we may see most clearly the effects and tendencies of such application in changing customs and institutions ; especially we see there the breaking down of the family as the foundation of morals and of the State, we see the detachment of the population from all local ties and traditional loyalties, and we see the processes of substitution of one stock by another far advanced ; we see many other social changes rapidly going on ; and most of these are features of the attempt to live by the light of reason in place of tradition, an attempt in part accepted as good and deliberately fostered, but in the main compelled as a consequence of the decay of tradition that inevitably results from the free mixing and blending of populations of very different traditions. For under such mixing and blending the various traditions weaken and destroy one another by mutual attrition. And not only are they weakened as influences moulding the characters of the citizens ; the very term " traditional " becomes a term of contempt and opprobrium ; and of all traditions the most delicate, subtle and indispensable, the moral tradition, is the first to fall into irremediable decay.

All this has its bright side, its advantages, and is welcomed enthusiastically by the very influential school of " behaviourists " who, having persuaded themselves and half their countrymen of the truth of a very crude and grossly over-simplified mechanistic view of human nature, have an overweening faith in the power of education (directed according to their own pathetically inadequate principles) to mould all men according to any desired patterns. The bright side is that such detachment from tradition renders men fit fodder for the industrial warfare and promotes the prosperity of the age of mass-production ; it makes them adaptable and amenable, subject to all the arts of the advertiser and the high-powered salesman ; it destroys their immunity to the rhetoric of the faddist, and renders possible rapid innovations, especially all such as superficially seem rational. It has rendered possible the immensely rapid spread of high-school and college education, as well as the growth on an astonishing scale of new fads, cults, superstitions and religions. If, as is too lightly assumed, the new, because consciously modelled and accepted, were always or generally superior to the old institutions (representing the traditional wisdom of the ages slowly wrought out by selection and long experience) such changes might be worthy of the enthusiasm of the radicals. But unfortunately we are still far too ignorant of human nature and the conditions that make for its best development ; we cannot safely discard the traditional wisdom until we shall have something solid to put in its place. The disastrous effects of the prohibition amendment furnish but one, though the most striking, illustration of these simple truths.

But the dark side of the picture is the enormous prevalence of corruption and of crime. Corruption pervades every level of society from the highest levels downwards ; and crime, waxing fat and prosperous, uses corruption in the spheres of business and of politics, in the judiciary and the police, to establish itself in impregnable defiance of society. So far has this gone that the very existence of civilized society is threatened. President Hoover has said : " Life and property in the United States are less safe than in any other country of the world. . . . I am wondering whether the time has not come to realize that we are confronted with a national necessity of the first degree ; that we are not suffering from an ephemeral crime wave, but from a subsidence of our foundations." He said also : " The most malign of all these dangers to-day is disregard and disobedience of law . . . our whole system of self-government will crumble either if officials elect what laws they will enforce or citizens elect what laws they will support." And that is just what both officials and citizens are doing on a grand scale. Chief Justice Taft has said : " Our criminal justice is a disgrace to civilization." And Mr. J. T. Adams writes : " We in America to-day are without the pale of this respect for law which is one of the fundamentals of civilization." Crime alone, apart from corruption, is estimated to cost the American nation three billion dollars annually, a serious offset to prosperity. And, in the main, these crimes are not the expression of dire need, of poverty and economic pressure ; the words of Presidents Taft and Hoover were uttered while America was at the peak of her prosperity. Crime and corruption on so vast a scale imply rather, as President Hoover says, the subsidence of the foundations of civilization.

My thesis, that the fundamental factor in this grave situation is the decay of traditions, is borne out by the well-established fact that the disorderly and criminal part of the population finds its recruits in largest proportion among, not the immigrants, but the children of immigrants. For it is these children who suffer in highest degree that weakening and deprivation of traditional influences which is the root of the trouble.

As regards the substitution of one human stock by others, America again affords the most striking illustration. Since the end of the civil war the number of immigrants from various European countries other than the motherlands of the original White Americans is estimated at some thirty millions. Yet the best authorities are agreed that, in all probability, the number of the population of the United States would stand at about its present figure if none of these immigrants had arrived. In other words, room in the social system for these millions of newcomers of alien stock and traditions has been made by restriction of the reproduction of the old American stock. It is probable that the introduction of the Negro stock, multiplied now to some twelve millions, has had a similar substitutive effect and will continue to exert such effect in increasing degree as the Negro stock becomes more completely acclimatized and subjected increasingly to the influence of hygienic measures enforced by States and municipalities. And there can be no doubt that, if no severe restriction had been imposed on immigration from Asia, the substitution process would have already gone very far in all the States of the Pacific coast and, perhaps, throughout the country.

Our most approved political institutions are much blown upon ; representative democracy based on universal suffrage, the ideal for which the pioneers of the nineteenth century enthusiastically strove, is no sooner realized than it proves so disappointing and inefficient that we see great areas reverting to various forms of tyranny.

Here America cannot claim to be in the van of progress ! The palm goes to Italy. But the decay of democratic parliamentary government in America is, though less advanced, more significant ; just because the development of democratic institutions and of civil liberty resting upon them has been the proudest boast, the very *raison d'etre* and highest spiritual aspiration, of America, the justification of her rebellion and of her disastrous civil war in which the flower of her manhood was wiped away and the nation left for ever and irremediably poorer in respect of its greatest treasure. It is this that gives so great significance to the breakdown of the two-party system, to the general contempt for legislatures and the legislative process, to the putting in power of such an administration as that of President Harding, the scandalous nature of which is fully known to few Europeans and indeed to but few Americans. And, above all, it makes it necessary to accept as a very grave warning the decay of civil liberty in America, the fact that, in the Land of Freedom, freedom is no more, a fact openly and deeply deplored by every intelligent American. Readers who may think these words too strong I would refer to the many essays dealing with this topic in the recent volume *Behold America !* [1]—all written by Americans.

[1] N.Y., 1931. Edited by S. Schmalhausen.

The decay of democratic institutions and the passing of freedom are the natural correlatives of the general decline of respect for law and the immense development of crime and corruption. They have already gone so far that it may well be questioned whether there is any hope of the survival of democratic institutions in America ; whether some form of Fascism or oligarchy does not offer the only hope of order and of the modest degree of freedom which is compatible with such a system.

Our so-called international law has been proved by the Great War and found to be an ineffective sham ; and the League of Nations vainly strives to bring about disarmament on the very questionable assumption that general disarmament, if it were possible (which it is not), would prevent war. Meanwhile we spend vast sums in preparation for war, sacrificing in the process the lives of our finest young men ; and we drift towards new wars that threaten to make an end of us and all our works.

I have examined this question at some length in my *Janus, the Conquest of War* (1927), pointing out, as a simple deduction from irrefutable facts, that in the absence of international police-force, disarmament is neither possible nor desirable and would not be effective in the prevention of war ; and that still less would a mere reduction of armaments be an effective preventive of war, however desirable on economic grounds. The course of events since that little book was published bears out my contention. The European powers have continued to make a great parade of preparation for a reduction-of-armaments-conference ; the Press and the public accept this sop to public opinion. All parties either blandly assert or assume that reduction of armaments will prevent war. No one stays to examine this ill-founded assumption. Yet surely it is obvious that, if two men each armed with two clubs are likely to engage in combat, you will not, by taking one club from each, appreciably postpone or diminish the probability of violence. And the same is true of nations. Yet year after year we continue to haggle and manœuvre in preparation for a conference for the reduction of armaments, as though such reduction were a sure guarantee of peace ; and gradually, as the date for the final effort draws nearer, the futility of the whole procedure begins to force itself into general recognition in spite of all our rationalizations.

The recent air-manœuvres over London have once more demonstrated, we are told by the Press, the impossibility of putting up any effective defence against sudden attack in force from the air. A leading newspaper [1] comments as follows : " If ' Redland ' and ' Blueland ' had been real States engaged in real war, how much would now be left of ' Redland's ' capital ? . . . Thirteen out of twenty-four daylight raids and no fewer than fifty-one out of sixty-eight night-bombing machines got through to London unchallenged. In air-war the advantage is overwhelmingly with the offensive, nor is this advantage confined to one side. The first week of a war between two neighbouring powers would produce no great battle between combatant forces, but it would assuredly compile an appalling record of wasted cities on each side of the frontier. The time is upon us when civilization must guard itself against self-destruction through its own achievements. . . . It is for mankind to say whether the conquest of the air is to be

[1] *The Observer*, 26th July, 1931.

40

judged as a fatal error or a supreme triumph." Does the editorial then proceed to consider how the suicide of our civilization may be prevented ? Not at all ; it goes on cheerfully and even enthusiastically to suggest that at an early date the London mob will be able to enjoy its Bank Holiday in Venice. That is typical of the attitude of the Press and the public towards this menace. The probability is too horrible to be calmly contemplated; and, seeing no means of prevention, we, with one accord, refuse to think about it and continue to make the agreed but futile preparations for the conference on the reduction of armaments. Is it not obvious that there is one and only one possible preventive of such an appalling disaster as *The Observer* coolly points to as the immediate result of the outbreak of European War— a result no less certain if armaments be halved, or even more drastically reduced ? This one preventive is the sure threat of instant and overwhelming reprisal for unauthorised attack from the air. And the only way such preventive threat (a threat that would never be converted into action, if only it were sure) can be made effective is by the maintenance of an international air-force with strict limitation of national air-forces. All this was set out in my *Janus* in 1927; and I have taken pains to bring it to the notice of a number of influential persons; yet, so far as I can ascertain, only one person in the world (namely Dean Inge) has had a good word to say for this proposal. It is not that anyone has examined the reasoning that leads to the conclusion and shown it to be in any respect faulty, either in form or in its premises. It is rather that the reasoning is in psychological terms, and though such terms are all-important in all political and economic discussion, our statesmen, our experts and our Press are only just beginning to take such terms seriously and are wholly unpracticed in the art of using them with confidence and precision.

The churches keep crying aloud their old stories and their old exhortations, but the people heed them less and less. In education we are all at odds; the only ideal that seems to make an effective appeal is that of keeping the children in school a year or two longer in order to make more jobs for their elders.

In the economic sphere the tragic absurdity of our predicament reaches its climax. Through the aid of physical science our powers of production have reached a very high point of efficiency ; an efficiency such that, if the whole machinery of production could be set working at full speed, every human being might be lapped in luxury of the most elaborate kind at the cost of a modest expenditure of human energy. Yet the whole world is poverty-stricken in various degrees, and even in America there are said to be at least six (perhaps eight or more) million workers out of work and a considerably larger number suffering serious deprivations.

If it be said that this is the nemesis of capital so confidently foretold by our Socialist friends, we cannot forget that in various countries the Socialists have attained to political power and have shown themselves unable to effect any remedy. While in Russia the Soviets are making a vast experiment in communism which, however successful it may prove in the purely economic sphere, seems likely to be put through only at a frightful cost of suffering and servitude, a cost that may well prove excessive and disastrous.

We live, then, in an age of grave social

disorder and threatening chaos, and it is in the main due to Science. What then is the remedy? The remedy for science is more science, more knowledge systematically organized. But what sort of science? Physical science has been the main agent in producing our chaos; and physical science can bring no remedy. Suppose that physical science should continue its brilliant and accelerating course; suppose that it should discover heavenly bodies a million times more remote from us than any yet observed; that it should enable us to see and hear at any moment what is going on at any point of the earth's surface and to travel thither in a few seconds; suppose that it should put at each man's service (on the average and in principle) energy equal to that of ten thousand slaves, instead of only fifty as at present[1]; suppose it should invent explosives and gasses a million times more destructive than any now available; suppose that another Einstein should convince us all that space is zigzag or that time is square. Should we be any happier or safer for any or all of these advances? Consider what would happen if some brilliant physical discovery were to put

[1] It is estimated that in America this number is about 150.

us in free communication with inhabitants of Mars not very unlike ourselves. We should forthwith be absorbed in efforts to prepare some ray with which to blast them from their planet, in the " purely defensive " warfare which our fearful imaginations would anticipate. And if some physicist were to realize the brightest dream of his kind and teach us to unlock the energy within the atom, the whole race of man would live under the threat of sudden destruction, through the malevolence of some cynic, the inadvertence of some optimist, or the benevolence of some pessimist. I submit that no such discoveries, nor any others that physical science could possibly make, could avail to remedy our condition. I would go further and assert that every step of progress physical science may make in the near future can only add to our dangers and perplexities ; for every step of such progress must increase the top-heaviness and the lop-sidedness which are the radical faults of our civilization.

I have no wish to belittle the achievements of physical science. They are immense and altogether admirable. I am concerned only to bring home to the minds of my readers the indisputable fact that the very successes of physical science, leading as they have done and inevitably must do, to rapid and violent changes in all our modes of living, are producing a state of affairs that is ever more unstable and dangerous, which urgently needs some large scale corrective such as physical science, no matter how successful, is, in the nature of things, unable to supply.

One of the great debts we owe to physical science in the main is the enlargement of our outlook upon the universe, our liberation from the old geocentric and anthropocentric cosmogony. But here, it seems to me, its great work is already accomplished. It is dangerous to prophesy, but it does seem very unlikely that physical science can render us any further important service of this kind. We are already sufficiently humbled before the immensities of Time and Space and Matter. When we are told that a star is one light-year distant from us, the statement far outruns our comprehension, and the revelation of new galaxies at distances hundreds and thousands of times greater leaves us unmoved. Even some finality of opinion among physicists as to the ultimate fate of the solar system would be of no crucial importance, and would affect our actions and our feelings but little. Most of us, I fancy, are not vividly interested in the question whether we are to be finally exterminated by fire or by frost. We derive more intellectual stimulus from the rapid succession of new theories of such matters, new theories of the ultimate fate of the stars, new theories of the atom, new theories of energy and its transformations, than we could do from any finality of doctrine; and such stimulus seems to be now and henceforth the major contribution of purely physical research to human welfare. Yet at very large expense we prepare to double the size of our giant telescopes and to drag them to the summits of yet higher mountains. This, in the present state of the world, may be likened to fiddling while Rome burns.

But it is in the sphere of material benefits conferred that physical science makes its largest claims on our gratitude. Here again America gives us the most vivid picture of the tendencies of our civilization. America more than any other country has benefited materially and economically from the discoveries made by physical science ; for she has carried to a further point than any other country the applications of physical science to production of commodities and to increasing the facilities of daily living. And, until the coming of the present economic depression, she was regarded with envy by all the world as the land of golden prosperity, of a high standard of living for all her people, of universal comfort and security. In a few years she immensely added to the total wealth of the country, and, from being a debtor nation, had become a creditor to other nations on a vast scale, and had far surpassed all her rivals in the international economic rivalry.

But what did all this mean for the common man in America, for the great mass of her people. Mr. Stuart Chase has recently told us, painting a clear and simple picture.[1] And the picture, to those who have been led by the glowing accounts of American prosperity to imagine all her people rolling in luxury, is disillusioning in the extreme. The farmers, still the backbone of the country, six million strong,

[1] *Prosperity, Fact or Myth*, N.Y., 1929.

have had little share in the general prosperity. Their position has become increasingly difficult ; one million have been driven to seek other occupations. The whole agricultural industry has made less, by 1,717 million dollars, than would suffice to pay each farmer the very modest wage of a little over £100 a year and four and a half per cent interest on his capital. " Relative to the rest of us, farmers have lost 17 points in the last ten years." That was in 1929 before the slump; and now their position, in view of the fall in wheat prices, is, in a very large number of cases, desperate.

As regards industrial workers, the general conclusion in 1929 was " That the only real prosperity has come in the ranks of certain skilled workers (the small group of the highly organized)—notably in the construction industry. Even this is relative." Whether or no real wages have risen during the prosperity period remains a disputed question. The best opinion seems to incline to the view that on the whole there has been a slight rise.

During the same period, thanks mainly to physical research and invention, the productivity of production processes has increased by leaps and bounds. "A group of thirty-five companies studied by Mr. Dennison show an increase in output per man of 74 per cent since 1919," and "The average factory worker in 1927 produced 40 per cent more by weight than he did in 1921." The result has been immense profits for the capitalist class; so that "just over 11,000 Americans have 100,000 dollars or more to spend a year, while 280 citizens at the upper reaches of the Pyramid carry on as best they may with an average of a little more than 2,000,000 dollars a year." Such magnificent results of industrial engineering naturally engendered the demand that the business-men and engineers should not merely run the country unofficially, but should be given official charge of its affairs. Hence a business-engineer was elected President; and then came the slump, and eight million workers were made workless and wageless. Surely an impressive demonstration of the inadequacy of business-engineering to the affairs of a nation! Some-one had blundered; but what the blunder was is still a matter of acute controversy.

Let us, however, note this one contributory factor, now generally recognized as a blunder. The business men, having greatly increased productivity, were faced with the need of larger markets. They turned their attention to salesmanship, and, with the aid of long credits to the ultimate purchaser pushed to the limit, achieved very large results. And this, it is generally agreed, has been no small factor in accentuating the present depression. If a sure method of accentuating business-cycles into violent booms and depressions had been desired, none better could have been devised. And this feature has this special interest. It was achieved, not by any application of physical discoveries, but rather by deliberate application of psychological principles. It served to show that immense results can be achieved by such application. But it was an application fundamentally unsound and unwise ; very successful in its immediate results, ultimately disastrous economically and socially, and involving an enormous overhead expenditure that absorbed a large part of the profits of improved productivity. Hear Mr. Stuart Chase again :

" While production costs are going down, distribution costs are going up. More and more people are being employed in the great American vocation of finding markets and shattering sales resistance. Mr. Julius Klein, of the U.S. Department of Commerce, computes a waste of distribution methods totalling 10 billions a year. He calls it the gravest issue now before the industrial community. His bill of charges includes excessive expenditures in sales promotion, disorderly marketing, careless procedure in retail trade, unsystematic warehousing, extravagant delivery service, ill-judged advertising, unwise instalment methods " ; to which sum of waste must be added, in any sane view, an enormous expenditure upon more or less successful advertising.

Another great item must be set against the success of the efforts to improve production, namely unemployment. Putting aside as abnormal the present very high figure of unemployment (some 8,000,000), let us note the incidence of this scourge during America's period of maximum prosperity. Says Mr. Chase : " After mechanized warfare, the bitterest thing in modern life is unemployment. Wars come and go. Unemployment goes on for ever. In the depression of 1921, some 5 millions were walking the streets, as factories and offices slowed down. In the spring of 1928, with business prosperity in full cry, it was estimated that 4 millions were out of work. To-day (September, 1929), with production roaring at its maximum, there are at least 2 million jobless, and quite possibly 3 million. A recent census of New York working class families showed 17 per cent without jobs at the time of the investigation . . . The new science of management has cut down labour, turnover, and thus unemployment, in certain individual plants. It has done nothing whatever about unemployment as a national problem. Nothing, that is, in a helpful sense. Its only contribution to date has been in the negative direction, to make unemployment worse . . . ' Technological unemployment,' as the phrase is now used, *means an ever-growing army* ; a total firing rate greater than a total living rate ; a displacement of workers by the technical arts *faster* than they can be re-absorbed in other occupations. Management fosters the technical arts ; management is accordingly the chief creator of technological unemployment, thus the central fire of productivity and commercial prosperity promises to wreck and destroy the hope, the happiness, the very lives of uncounted human beings, in an ever widening arc." [1] To which must be added that the term " management " as here used means in the main the application of physical science.

Mr. Chase tells us that " management " has been immensely successful in speeding up production-processes and reducing their costs, but : " It is not without significance, however, that only 20 per cent of the officials interviewed (in an extensive inquiry) regarded the bettering of the state and morale of their employees as an aim in progress, while 80 per cent concentrated on bigger buildings and faster machines. The human aspect is not as yet a leading aspect of the new science of management."

Summing up on American prosperity of the period of maximum prosperity,

[1] Op. cit.

Mr. Chase writes : " If we were barely comfortable in 1922, we ought to be reasonably comfortable to-day. But, of course, the fact is that some 80 per cent of all American families lived below the budget of health and decency in 1922, and the 20 per cent increase in per capita income since that date, while it has helped to be sure, still leaves probably two-thirds of all families below the line . . . We have added a little real income and considerable fluff to the totally inadequate distribution of goods and services obtaining in 1922. Is this prosperity in the deeper sense ? No. The most that can be said is that the last seven or eight years have registered a rate of advance in the direction of a prosperity which may some day be achieved."

When the substance of this little book was delivered as the Mond Lecture, some of my friends told me that I sketched an unduly pessimistic caricature of our times, and twitted me on having joined the army of croakers. I venture to think that it is they who are mistaken. In England anyone returning from a long residence abroad is very favourably impressed by the evidences of great improvements in the condition of a great part of our population. But these improvements have been achieved by means of a wholesale process of living on our capital and running up a large bill for our descendants to pay. This process can be ultimately justified only if we can succeed in so ordering our social and economic and international affairs that the already existing possibilities of greatly increased productiveness can be realized in the fullest way, so that the total production of wealth may be greatly increased and the bills we are now running up paid without undue effort and ruinous cost. And it remains still to be seen whether we can achieve the required degree of such ordering. That is the problem which confronts us and on the solution of which all depends. My position is neither optimist nor pessimist. I am merely concerned to indicate the full magnitude and dffiiculty of the task, and the factors upon which its accomplishment depends.

But it is relevant to cite the opinions of some men of wide experience in public affairs. Putting aside such views as Oswald

Spengler's, which seem to me founded upon a false method, I cite the opinion of two such men, one a leading Conservative, the other an equally experienced and distinguished member of the Liberal party. In a posthumously published essay,[1] the late Duke of Northumberland wrote as follows : " It is surely evident that we are approaching a great crisis in world affairs. The hopes of ordered and continuous progress cherished hitherto, rest on no solid foundation ; the hopes of universal peace are a dream, for the method of attaining them offers no chance of success ; the hopes of continued development of political and social institutions are equally unattainable, because those institutions are bankrupt. Those who still pin their faith to them are falling into the same error as that with which they have previously charged the reactionary element in all countries, lack of imagination and failure to realize the spirit of the age . . . It is impossible to avoid the conclusion that the element of decay which we have observed in recent European history must eventually destroy the remaining institutions upon

[1] *National Review*, July, 1931.

which our society is based, because the foundations of those institutions, religion and nationality, have lost or are losing their force." To this I add a sentence from an article by Mr. A. Wigglesworth in the same review : " Finance, commerce, religion, custom, and the methods of life are all in the melting-pot, and we grapple in darkness."

Now the opinion of a leading representative of Liberalism, Professor Ramsay Muir, distinguished historian and active politician. He was recently reported in the Press as saying : " Civilization was on the verge of collapse, and the German crisis was a striking illustration of the danger. . . . If in the next few years there is not a substantial movement towards greater freedom of trade among the nations, and, still more, if Britain, of all countries, joins in this madness that is ruining the world, civilization will go with a crash from which there is very little chance of recovery." [1]

I may add that a person who is in touch with many sources of information that are closed to the public recently asserted in conversation that during a

[1] *The Times*, 19th July, 1931.

few days of the present summer (1931) our civilization tottered on the verge of total collapse. I am not at liberty to give his name. I may only say that he is very widely and highly regarded for his prophetic and generally optimistic views of world affairs.

In this appalling situation, in face of this dread prospect, we continue actively to augment the sources of our trouble ; we continue deliberately to increase this lop-sidedness of our civilization, we devote more and more of our resources to physical research. " Those whom the gods would destroy they first make mad."

It is only the biological and especially the social sciences founded on biology that can save us.

This sentence needs some expansion. My thesis is that in order to restore the balance of our civilization, in order to adjust our social, economic, and political life to the violent changes which physical science has directly and indirectly produced, we need to have far more knowledge (systematically ordered or scientific knowledge) of human nature and of the life of society than we yet have. First we need to know the truth about differences of fundamental constitution between races and individuals. Are these very slight and unimportant, as one large school of opinion confidently asserts? Or are they, though difficult to define, of profound importance and very difficult to modify, as others of us believe, though at present we cannot adduce conclusive evidence? Is it true that some existing human stocks are far more capable than others of producing and maintaining a high civilization? Can we hope that, with or without socially directed effort, the existing races of man are likely to advance in respect of the qualities that make for a high level of civilization? Or is a general decay or falling off in quality probable or inevitable? What measures can be taken to promote the one possibility and make the other more remote? Are we at present covering over, by means of improved hygiene, education, and general conditions of living, a subtle degenerative process affecting perhaps all the more civilized part of mankind? May improved conditions of life, with improved training and efforts at self-improvement, effect improvement of the race, or secure it against deterioration, as the Lamarckian theory, if it could be substantiated, would lead us to believe? What are the effects of the cross-breeding of the various human stocks? Are they all good or all bad? Or are some good and some bad; and if so which? All these are questions profoundly important for the future of mankind, to which biology at present can give no sure answers. Briefly, then, we urgently need a well-founded theory of evolution such as at present we lack; and we need knowledge of its detailed application to the human race.

Secondly, we need the development of the Social Sciences, economics, politics, jurisprudence, criminology, penology, history, social anthropology, and all the rest, for our guidance in all social and political problems, in face of all of which we stumble blindly along amidst a chaos of conflicting opinions. And all of these need for their foundation some sure knowledge of the constitution of human nature and of the principles of its development; in other words, a sound psychology. Take the great question of perennial dispute—the most desirable political-economic organization of peoples; should it be democratic

or authoritarian ?   Should it be in-
dividualistic, socialistic, or communistic ?
The answers to all such questions depend
upon the assumptions we make (and in
the absence of sure knowledge, are free
to make) about human nature. Will the
citizens do their duty and lead the strenuous
and co-operative life without the spur of com-
petition and the threat of economic misery ?
No one knows, and we all make (for the
most part implicitly only) what assumptions
we please, and shape our answers and our
political applications accordingly. Many
years ago I had the pleasure of hearing
an eloquent address from Mr. Philip
Snowden to the members of the Oxford
Union.   Mr. Snowden sketched his
socialistic paradise, assuming on the part
of the citizens of that paradise the
spontaneous display of all the fundamental
virtues.   Then turned upon his hearers
with the challenge—" Will it work ? If
you doubt it will work as I say, it can only
be because you are cynics and take a low
view of human nature ? " What more could
be said in the way of argument ? The
discussion, as all such discussions at
present, proceeds from premises which

are but vague opinions and ill-founded assumptions, determined mainly by temperament and the accidents of personal experience, of luck and ill-luck.[1]

[1] It is noteworthy that Socialists are advancing beyond this crude position, and, if we may regard Mr. Norman Thomas as a representative Socialist, they may even be said to be showing the true path to the social scientists in general. In his recent book* he writes, in reference to the cry "You can't change human nature": "But we do not dispose of the case against human nature by proving how much of it critics possess. Really to deal with the problems involved would require not only an expert knowledge of biology, psychology, anthropology, and history which I do not lay claim, but also, I suspect considerably more progress in these sciences themselves, especially among our warring psychologists. Nevertheless, we shall have to face some of the charges against human nature and work out some modest and tentative conclusions as a condition of hopeful effort. The available evidence does not bear out the familiar contention that we are condemned to the capitalist system for ever because men will work only for profit. Behind this contention there has been far more assertion than proof. Indeed that judgment must be passed on all *a priori* statements of the hopelessness of expecting men to win a better world."

And on another page he writes: "Reference to that blessed word 'psychology', in which our generation is perhaps abnormally interested, calls attention to the absence of any adequate concern among Marxists over psychological problems that so agitate us. Even if Dietzgen, the so-called

* *America's Way Out*, N.Y., 1931.

Consider the sphere of international relations. Nationalism, it is generally agreed, has been the greatest moulding force in the history of the nineteenth century; and it has become acutely accentuated in recent years. It is widely denounced as the greatest evil of the present time. Yet others, like the late Duke of Northumberland, deplore the decay of nationality and of patriotism; and where, as in India and China, nationality is as yet merely an ardent

philosopher of socialism, is included and given greater weight in its councils than he had or indeed deserved, Marxian Socialists, themselves, like their capitalist contemporaries, never really faced such absorbing questions as these. If economic conditions determine men's ethics and politics, how is the job accomplished? How do men think? What place has thought or reason in guiding men's social behaviour? Why do men as individuals and in groups so easily miss their own real interest? Why do we tend so easily to stampede in crowds? Why is a crowd loyalty, like absolute nationalism, so persistent even when it is clearly out of line with world-wide economic trends? Is human nature capable of the degree of intelligent co-operation which the attainment and perservation of Socialism will require?"

All of which shows that Mr. Thomas the Socialist sees clearly—as so many of the social scientists do not see—that answers to all our social questions presuppose a well-founded psychology.

aspiration, the very persons who most loudly denounce nationalism in Europe as a child of the devil exhort us to encourage and sympathize with the nascent but profoundly disturbing nationalism of these ancient civilizations. And, again, when the late President Wilson proclaimed the great principle of self-determination for all peoples and succeeded in applying it so far as to Balkanize half of Europe, his supporters and accomplices were in the main the same persons who now so loudly denounce nationalism. What, then, is this greatest of all modern forces, nationalism ? And what are patriotism and nationality, and what their relations to nationalism ? These are questions that must receive clear answers before we can hope to emerge from this welter of confusions and con- tradictions. And all these are psychological questions ? Whatever else this much discussed force, nationalism, may be, it is in some sense a component of a multitude of energies that reside and operate in the breasts of human individuals ; and its workings cannot be understood until we have at our command some well-founded psychology, both individual and collective.

Again, economics is, or should be, a science that deals with the desires of men, their strivings and their intellectual operations, the processes by means of which they strive to attain satisfaction of those desires—primarily and fundamentally a psychological science. Yet one half its official exponents deny this simple truth, while the other half pays it lip-service only. But of this more later.

Or consider jurisprudence and the allied problems of criminology and penology. It is not true that all these complex blends of science and philosophy are concerned to regulate, to direct, to order and reform the conduct, the thoughts, the intentions, the feelings and the actions of men and societies? And how shall they do this unless they truly conceive, however inadequately in detail and in special cases, the general principles of human motivation. It is pathetic to read the works of a great jurist like the late Sir James Stephens, to see this powerful intellect struggling vainly with its complex problems ; vainly, because he starts out with a fundamental assumption about human motivation, the assumption of hedonism, widely current in his time, yet radically false and misleading. It is like watching a lion struggling in the toils of a strong net.

What, then, of psychology itself, the " science " that should be and some day must be the consciously accepted foundation of all the social sciences ?

It remains a chaos of dogmas and opinions diametrically opposed, a jangle of discordant schools and sects ; a field exploited by quacks and charlatans of every sort, preying upon the ignorance of a deeply interested public which knows not which way to turn for authoritative guidance. This is not merely because psychology is the most difficult of the sciences ; nor altogether the consequence of the uncertainty of its biological foundations. The chaos is largely due to the neglect of the field by our universities and is likely to continue until some time after they shall have fully accepted their responsibility in the matter. The proper study of mankind is man ; on this we are all agreed. But, though our leading universities have long devoted themselves to the study of man's works, they have been very slow to make any effort to study man himself. Until very recently there have been no chairs of Anthropology and of Psychology in the British Empire ; and our neglect to take advantage of the unequalled opportunities offered by its multitude of peoples of all grades of culture will for ever remain a serious reproach to us. It is a fair presumption that, if we had not grossly neglected these studies, we might have avoided the present difficulties in India and China ; and our Indian Empire might not have been lost. Even at the present day the number of university chairs in these two subjects together is only half a dozen, and nowhere in the British Empire is there a department of either subject with any but most scanty and inadequate equipment and personnel. Thanks to the energy and munificence of Dr. C. S. Myers, the National Institute of Industrial Psychology has secured recognition by the industrial world and begun to show how much may be done in that very special field. But, apart from that, British activity in these all-important fields makes a lamentably small showing ; and psychiatry might almost be said to be non-existent, while the patients in our mental hospitals grow steadily more numerous, a costly burden, likely to swell constantly until such time as we can learn to effect some better adjustment between ourselves and the conditions of our lives, so violently disturbed and tremendously complicated by physical science. Research on the chemistry of aniline dyes is considered to be of prime importance and worthy of ungrudging support, because it puts money into the pockets of a few. But are we any better or more beautifully clothed or better housed for all our aniline dyes? Though men decay and nations collapse, let us have our aniline dyes and further research upon them ; for it pays—in the narrowest sense.[1] Our present tendency is towards a world of gaudily attired neurotics and maniacs housed in barracks where they will pass the time between crises and disasters pleasantly enough, pressing innumerable buttons to set in automatic action the inane products of jazz and movie factories.

I have said that at present we have no such sciences. The statement may seem extreme.

---

[1] I am not unaware of the important relations of dye-stuffs to explosives and poison-gases.

It is true we have some beginnings of biology ; notably a vast mass of " data ", some highly specialized departmental studies, and some successes in the sphere of empirical medicine. But we have no biology that can serve as the basis of the social sciences we so urgently need. And the social sciences themselves are merely a faulty sketch of a programme.

I cannot pretend to examine each of them in turn. Let us glance very briefly at that one which has the longest history and the most considerable body of students, and which has the most immediate bearings on social practice. I mean Economics. Will any one affirm that Economics is a science ; that it is anything more than a frightful mess of statistics and highly questionable theories ? The supreme test of a science is its power of prediction. It would be too much to say that the predictions of the economists are always wrong. But, if some of them have not been wrong, is it not merely because, among a multitude of predictions, some must hit the mark according to the law of chance ?

Apply the less exacting test of power to explain after the event, and what do we find ? A sheer chaos of opinions, the highest authorities directly opposed to one another and a multitude of smaller fry with their own versions. Take the question of the grounds of the present world-wide economic slump, or almost any other economic phenomenon you please ; and it is always the same. Could there be a greater diversity of opinions and acuter opposition of authorities ? Take the single question—How far is the distribution of gold a principle factor ? And when we come to recommendations for economic betterment we find chaos raised to the nth power.

Consider the chaos of opinions and recommendations concerning the gold-basis. The necessity of a gold-basis has been for generations an accepted dogma. Only a few of us have wondered whether it was not merely a superstition. And now at last it is blown upon from many quarters. Economists of the highest repute tell us that it is not only a superstition but a most pernicious one, the main ground of all the economic sickness of the world in general and of Great Britain especially. A leading journal [1] announces : " A gold reserve against currency serves no useful purpose—a few years ago most bankers and financiers would have been aghast at such a doctrine ; but now there will be few to dispute its theoretical

---

[1] *New Statesman*, May, 1931.

soundness." Consider the controversy over deflation and inflation. Does it not justify the adjective "chaotic"? Can such a chaos of opinion be called Science?

Lord D'Abernon is recently reported as saying : "An infinity of views showed a total absence of clear explanation or of definite guidance. A systematic regulation of value of gold measured in commodities is a paramount necessity." Among others, Professor Gustav Cassel and Mr. Reginald McKenna seem agreed in holding that the deflation (made for the sake of maintaining the gold-basis) has been a main factor in bringing about the present economic state of Great Britain. But on the question whether this was worth while they are acutely divided. Professor Sprague, the American economic adviser to the Bank of England, is reported to have said recently : "Although statisticians and economists in recent years have given more attention to the study of trade fluctuations than to any other problem in the wide field of economic inquiry, there is a conspicuous absence of agreement in our conclusions regarding the causes of fluctuation and the means of establishing a closer approach to stability in the conduct of economic activities." Disagreeing with Lord D'Abernon (the two utterances were published in parallel columns of *The Times*, 17th June, 1931), he proceeds to annihilate those many economists who, laying the blame on bankers and on financial policy, call for increased supplies of money and credit, and to advocate what he calls " the equilibrium view ". " According to this view banking and other financial agencies cannot bring about a recovery from a major depression. The economic adjustments that are required can neither be imposed upon the community by means of limitless contraction nor induced by unlimited credit expansion. Other agencies are required."

About the same date (9th June) another great economic authority, Sir William Beveridge, gave a nationally broadcast talk on the same general topic. The essential feature of the depression is, he said, the fall of prices ; and that in turn is due to restriction of credit ; but he refrained from all attempt to say how and why credit is restricted, contenting himself with expression of the pious hope that some day financiers will be able to understand the ebb and flow of credit and to control it. Incidentally, he asserted that when prices begin to fall, the fall is accentuated and prolonged because people in general abstain from buying commodities in the hope of a further fall. This statement has at least the merit of taking account of the so much complained of human factor. It is a frankly psychological proposition. But is it true? Is it based on wide inquiry into the motives of buyers? I doubt it. Would it not be at least as plausible to assert that when prices fall the buyers rush in to seize the opportunity. Are buyers in general so canny as Sir William asserts? The direct opposition of economic views is the more striking when each of the opposed views is expressed in concert by a large group of experts. I note, as an example, that the *Report of the Liberal Industrial Inquiry* recommended restriction of foreign investments; while, with the same end in view, namely the re-establishment of British Industries, the Balfour Committee made the opposite recommendation.

Occasionally we do find a consensus of

economic opinion. Up to the outbreak of the Great War it was, I believe, the common opinion of all economists that a great European war could last only a few months at most : for, it was agreed, such a war must quickly be brought to an end by the economic exhaustion of one or both parties. The economists had, as usual, neglected to take account of the most important factor concerned ; they had ignored the probability, quickly realized on the outbreak of war, that the motivation of almost all persons concerned in the economic activities of all the nations would be profoundly changed. It was these changes that made possible the immense economic, financial, and industrial efforts, efforts which enabled the prolongation of the war for more than four years, although the war assumed vaster proportions and involved far greater and more extensive economic waste than anyone had foreseen.

In every branch it is the same story. The books on economics are full of iron laws, inexorable laws. Yet in any true sense is there an economic law anywhere in sight ? The very language the writers use is hopelessly loose and confused. Let me cite one instance which points to the root of the trouble. Arnold Toynbee, comparing economics with physics, wrote : " But the economist has to deal with facts which are far more complicated, which are obscured by human passions and interests, and, what is still more to the point, which are perpetually in motion." What does he mean by saying that the physical facts are at rest and the economic facts in perpetual motion ? We can only guess. But more significant is the statement that the economic facts are obscured by human passions and interests. He should have said rather that the essential economic facts are " human passions and interests ". The economist constantly speaks as though there were a realm of facts and laws which he might reduce to order, if only there were no human passions and interests to pervert his facts and laws and to frustrate his praiseworthy efforts to reveal the facts and formulate the laws. Toynbee is out-of-date, you will say. But when I turn to the most authoritative contemporaries I find the same state of affairs. Mr. A. Loveday, the economic adviser to the League of Nations, has recently published a book. In that book he frequently writes of economic " forces and tendencies " that are said to operate with greater or less force in this or that country. It seems clear that when statistics reveal any continuing change, he postulates a " force and tendency " as the cause of that change. But what the force and tendency may be he does not stop to inquire. He, like Toynbee, merely querulously implies that human passions and interests are unwelcome disturbing factors which make difficult the economist's task of defining these " forces and tendencies ". He writes : " I tried to lay bare certain forces and tendencies which owe their existence mainly not to conscious and concerted human effort, but to unco-ordinated changes in individual action and to the natural growth of wealth." He speaks of various forces and tendencies being checked and arrested, as though they still continued in being, while counteracted by others, in analogous fashion to physical forces.

I turn to Sir Josiah Stamp, and I am happy to find that he insists, as I am doing, on our urgent need for economic science. " We are in real peril, and a serious breakdown of our economic society is far from being impossible." " To-day is the day when of all times the drive is wanted in Economics . . . we cannot have too many hard and brilliant thinkers in that field, too much money poured into research, and too much patient self-denying effort to advance the science." And he rightly points out that mere accumulation of statistical facts is of no avail. " We have vast masses of facts . . . in the coal industry in England every fact has been meticulously known . . . no industry . . . is more fully documented than the coal industry, but unfortunately the psychological conditions behind it . . . still have their influence on it." Here you see again the same assumption of a realm of economic facts and wickedly disturbing " human passions and interests " frustrating the good intentions of the economist to make a science ; the assumption of laws that *might be* iron and inexorable if only human passions and interests would not interfere. And the same old fundamental error of the economists comes out, when in other essays the same authority repeatedly distinguishes between, on the one hand, economic facts, and, on the other hand, our thinking about them and our personal attitudes towards them. There by implication he denies that our thinking and personal attitudes are economic facts ; whereas they are in truth the most important of all economic facts. Economics deals with values, and there is nothing good or bad (nothing of value), but thinking makes it so. How absurd, then, this pretence that economics can abstract from the human factors and discover a realm of facts and laws that may be independent of this cruelly disturbing factor, human nature and its vagaries !

It may be said in defence of economics that it is now facing altogether new and unprecedented problems, and that economists have not yet had time to solve them. But, if we go back to the great doctrines discussed by economists throughout the last century, we find the same state of affairs, a chaos of opinions in regard to these corner-stones of "economic science". Take a single example, the famous " theory of marginal utility ", propounded more than a hundred years ago, accepted by many leading economists, and rejected by others. Of it Dr. O. Spann, the leading historian of economic thought, writes : " The fundamental notions upon which the doctrine of marginal utility has been built up are untenable." And : " Since the theory of marginal utility sets out from premises which are unsound both methodologically and in respect of their content, we are not surprised to find that it leads to erroneous conclusions." Again : " For all these reasons the theory of marginal utility has led only to the formulations of sterile pseudo-problems and to empty sophistications."

As another instance of one of the most approved " iron laws " of economics, consider the " law of supply and demand ". The big businesses of America have discovered that it is often possible to create demand to order, and on the largest scale. And, if this is not a sufficiently striking demonstration of the plastic nature of the " iron law ", consider the history of the Chinese and Indian boycotts of British goods. These surely were most impressive instances of the power of " human passions and interests " to interfere with the purity and rigidity of " economic laws ".

The assumption of an economic Robot dates from the early days of the classical Political Economy ; it still survives in the implicit assumption that the laws of economics would be valid if only men were such Robots.

It is the basic error of most economics ; the assumption of an iron man governed by iron laws ; a Robot so simple that it may be left out of account after we have made a few deductions about its working. It is no wonder that economics, with its iron laws true only of iron men, is often called the most inhuman of the sciences ; for economics strives constantly to ignore and abstract from the human factors which are the most important of the facts and which express the most important of the laws of which it must take account. Economics is surpassed in its inhumanity by its special branch, finance or banking. Do I need to illustrate the fact ? Perhaps it may suffice to refer again to the banker's superstition of the gold-basis. Or take a comment made by one of our great Dailies [1] on an international banker who has recently ceased from troubling, one who had played a large part in bringing about the present state of the world. It wrote that he knew everything about money but nothing about the imponderables. By that it clearly meant that he knew nothing about the human factors ; which was to say that all his opinions were worthless, probably, in many cases, far worse than worthless. For what is money apart from the human factors ? It is literally nothing. The value of money, like all our other values, is a function of human nature. What a shocking state of affairs ! That the welfare of many

[1] *Morning Post.*

millions of men throughout the world should be at the mercy of a man who knew all about money, and nothing about the imponderables. No wonder the world is in chaos !

The importance of this topic will justify a few citations further illustrating the chaos that is the science of economics and the root cause of that chaos, of the incompetence of the science to perform the functions of explanation, prediction and guidance expected of it. The citations express in the main criticisms made by influential observers of affairs who are not themselves professed economists.

Mr. Collinson Owen (*Sunday Pictorial*, 19th July, 1931), an experienced journalist who has made effective studies of American conditions, writes : " More and more we are beginning to realize that we do not understand the forces we are supposed to control. It is now only a month . . . since President Hoover made his proposal for a debt holiday for one year—and since then political and financial events have zigzagged up and down in the most unexpected ways, forming the sort of graph you might find on an engineer's office wall. Nothing has gone according to plan. That proposal, largely designed to save Germany from an immediate crisis, seemed to achieve that particular purpose almost at once. And then, within the past few days, Germany has slipped into a secondary crisis, even more extreme than anything envisaged a month ago. We are living in a crazy world, in which nothing can be foretold, and anything may happen to-morrow.

"The fact is that though the world's financiers find their own sphere of operations more than they can understand and provide for, except from day to day, we must not blame them too much for being out-distanced by their own science. They are not dealing only with pure finance, with no doubt the usual admixture of impure finance. Political, that is to say human, values are also involved, chancey factors which at any moment may switch the whole course of events into another channel and undo all sorts of good work done."

There again we note the common fallacy of the economist, the assumption that there is, or might be, a realm of *pure* finance, if only chancey unpredictable human nature would not interfere, if only men were financial robots. Whereas, of course, finance is, hardly less than war, a realm of human passions and interests and *nothing else.*

In an article on " The Evil Influence of Wall Street ",[1] Mr. G. T. Flynn, an expert critic, writes of " the mysterious and irresistible laws of finance " ; and of the capitalist system centred in Wall Street he asserts : " The simple truth is that not one disinterested student qualified to make the study has ever attempted an exhaustive examination of the social and industrial and financial reactions of the market operations." He goes on to write of some striking instances of false predictions by economists. He recites how, on an October day of the year 1929, Professor Irving Fisher, generally regarded as the most influential economist in America, " announced that the market was not inflated, stock prices were not too high ; ' in my opinion current predictions of heavy reactions affecting the general level of securities find little, if any, foundation

[1] The recently published volume *Behold America!* N.Y., 1931.

in fact.' *On the following day* the stock market broke and financial values to the extent of five billion dollars were wiped away." A little earlier the influential *Magazine of Wall Street* had made similar predictions. Says Mr. Flynn : " There were many economists of banks and large industrial corporations . . . who advised clients that all was well in a rose-coloured world. Almost every big business concern, *on the strength of promises of these pro-fessional prophets*, made preparations in the fall of last year for expanding markets. Within a moment the market fell and then the oracles saw what had been going on for some time." He sums up as follows : " From all this it may be seen, first that the experts denied that a slump was coming ; second, that a slump came ; and third, that all first aid methods were unavailing." Then Mr. Flynn points to various attempts of men to make economic laws work for them by means of trade agreements, by which he means agree-ment to keep up prices and restrict production. He points especially to copper, rubber, and coffee.

" Now these phenomena raise up a

group of problems which, I hoped, when the Wall Street crash advertised them so thoroughly, would be dealt with by the responsible industrial and banking leaders of the country. Apparently no one has given this a second thought. We see on every hand movements for reviving and extending price-fixing agreements and devices, our own Government attempting it in the form of a gigantic and futile experiment [1]—and so we go forward in the old round of sequences. We will get business under way, get people back to work, increase corporation profits, work up another speculative mania, produce more than we can use, use up our credit resources on speculation, cripple the building industry again, get business into a decline, bring on a stock market collapse, slide into a depression."

[1] A reference, presumably, to the recent attempt of the American government to bolster up the price of wheat. Referring to a similar attempt, *The Times* remarks: " Indeed, the various attempts of this nature which have in the past few years been applied to so many other important commodities besides cotton, such as wheat, copper, coffee, and sugar, to mention only a few, may be regarded as one of the principal causes directly responsible for the present world depression ".

Mr. Wyndham Lewis writes (in his recent book, *Hitler*, London, 1931): " Psychologically, if not physically, the peoples of the whole earth (whether it be China, Germany, Russia, Australia) have entered into the shadow of ' misery ' of mental misery; in the case of many nations, or portions of nations, of physical misery as well. Now, in the light of the belief I have been examining [the belief, called by him ' credit-crankery '] this ' misery ' is purely and absolutely artificial. It is the result not of an actual, a natural want, but of an artificially-fostered, sedulously-contrived want. Obviously there is no real want ; there is an enormous abundance of everything, if man's technical power to produce were made use of and put at the disposal of all. But for some reason or other we have slowly been con-ducted into such a state of affairs that, in the Lap of Plenty, we have agreed to starve. And the ' Science of Economics ', as usually practised, does certainly seem to be there merely in order to confuse us and to throw dust in our eyes."

Mr. T. S. Elliot writes : " I am not convinced that the accomplished economic specialists of the Harley Street of finance always know what they are about themselves. I have served my own apprenticeship in the City ; endeavoured to master the 'classics' of the subject ; have written (or compiled) articles on Foreign Exchange which occasionally met with approval from my superiors ; and I was never convinced that the authorities upon whom I drew or the expert public which I addressed understood the matter any better than I did myself—which is not at all." [1]

Our major prophet, Mr. H. G. Wells, writes, in his genial way, that the economists " have produced a literature 10,000 times as bulky, dreary, and foolish as all the outpourings of the medieval schoolmen."

If we look round to pick out the really notable contributors to this field, those who have done something to illuminate it, we find it is those who do not neglect the imponderables, who treat them, not as annoying and undesirable complications of some alleged purely economic facts, but as the most essential, the all-important, economic facts. Such was Walter Bagehot. Such in our own day is

[1] *The Criterion*, January, 1931.

Mr. J. M. Keynes. I look at a recent short article from his pen discussing Tariffs, and I find thirty references to the imponderables. And one of these imponderables is mentioned seven times, and would appear to be, in Mr. Keynes' view, the all-important factor in the contemporary economic world and the lack of it the prime factor in maintaining the distressful position of Great Britain, namely *confidence*.

Here we have a pointer; it points directly to the conclusion I am urging. What is this imponderable so overwhelmingly important in the economic sphere—this *confidence* ? Surely it must be worth while to study it ; to try to learn something of its nature and conditions and laws. For surely it is a natural phenomenon that has its conditions and conforms to laws. And the same may be urged of all the other imponderables mentioned by Mr. Keynes and of many others which are no less important for economics and for all the other social sciences.

Mr. Keynes is not alone in assigning to confidence a rôle of first-rate importance. Almost every article in the Press discussing economic and international problems during recent months lays similar stress on the same mysterious imponderable. The lack of it is deplored ; all men and nations are exhorted to possess it ; it is generally regarded as the *sine qua non* of peace, prosperity, and progress. But does Mr. Keynes or any other economist or political scientist propose to inquire into the nature and conditions of confidence ? Not at all ! Confidence remains an utterly vague word. It is clear that it stands for some natural phenomena of immense importance ; it is clear that the conditions under which confidence ebbs and flows are vastly complex. It is clear that the phenomena and their conditions are deserving of the most intense study, that, in fact, all economists and statesmen should be chiefly concerned to learn all that can be learnt concerning them. But no ; we spend millions on the search for new and deadly gases, and have tens of thousands of experts engaged in research for the improvement of motor-cars, but we spend not a penny on research into *confidence*, which is more vitally important for the very life of our civilization than all the multitudes of physical researches now going on in every civilized land ; more important than all that these researches can possibly reveal to us in the course of a century.

Suppose that some highly unorthodox economist were to suspect that psychology might have some bearing on his problems and were to ransack the shelves of a psychological library to find what psychologists have to say on *confidence*. He would hardly find one book or one author in a hundred that has anything to say on the subject. And why? Not because there is nothing to be learnt, but because in this country the study of such things has been grossly neglected, discouraged, and condemned ; while in America, where professors of psychology abound, they are ruled by the mechanistic dogmas dear to Professor Hogben ; and, in the intellectual atmosphere engendered by that rule, the word *confidence* is meaningless, and must remain so until such time as Messrs. Hogben and Needham or other bio-chemists shall have discovered the chemical formula of confidence or the electro-magnetic equation in which it may be expressed and a truly scientific knowledge of it built up. Let us concede the possibility of such discovery and the further possibility that a chemical antidote for lack of confidence may be discovered. The drug will then be manufactured on the largest scale and administered in full doses to all concerned in economic and political activities. The world will overflow with *confidence* ; all caution, all circumspection will be banished as by a magic wand, and the world will rush *confidently* to meet disasters and disorders compared with which those of the present time will seem trifles.

It is no part of my task in these pages to attempt to make good the deficiences indicated. Nevertheless, it seems worth while to point out that we are not condemned to remain for ever in total ignorance of this all-important imponderable, confidence. Even now, much might be said in illumination of its nature and conditions. We should have to begin by studying confidence in the individual ; and we should find that in every concrete instance we are concerned with a problem of human energetics, of some specific complex of impulses, desires, mutually co-operative or antagonistic, each having its own peculiar function in the constitution of the individual and its peculiar history in his development. We should find also that a multitude of diverse external influences are capable of promoting or of restricting confidence in each instance. Then we should have to go forward to a study of the still more complex conditions of group-confidence ; here we should encounter most of the major problems of collective psychology.

It is encouraging to find gleams of light breaking out where one might least expect to find them amid the prevailing darkness. In a weekly review published by a capitalistic financial company I find an article full of humane wisdom and sound psychology of a commonsense kind.[1] Citing Mr. Gerard Swope, president of the General Electric Co., the article proceeds as follows : " How demoralizing it is for a man who is able and willing to work, because of the breakdown of our social and economic machinery, not to be able to secure such work ! A human being laid off for lack of work, after the

[1] *The Bache Review*, published by G. G. S. Bache and Co., 16th May, 1931, N.Y.

harrowing experience of endeavouring to find work is never again as self-respecting and self-reliant a human being as he was before—he has lost something that can never be replaced. The demoralization of idleness and irregularity of employment is not only demoralizing to the individual himself but to his family, and reacts upon society. Such an employee, with that constant fear hanging over him of being laid off and unable to provide the necessities of life for his family and himself, approaches his work with a mind that is not free from worry. He approaches it as a task to be accomplished with the largest amount of compensation to be immediately secured.

" It is confidently believed that if the fear of lack of work were removed from the mind of the working man, he would approach his work with an entirely different attitude, his intelligence would be aroused, his interest stimulated, and he would have enthusiasm and satisfaction from the work to be done. As a result the quality of the workmanship would be better, he would feel free to suggest better methods, which would lower costs and be reflected in lower selling prices to the public [the bugbear of most economists just now] and assure a wider sphere of use for the article itself."

The article cites the equally interesting psychological reflections of another prominent business man, Mr. E. C. Vogel, Chairman of the Commercial Investment Trust Corporation : " Business is a result of human effort, its conduct is guided by man, who, unfortunately for himself, is not always calm or clear thinking. He suffers at times from exhilaration and exuberance and at other times from depression of spirits, and these temperamental and psychical changes influence his action in business. Prior to the stock market crash in October, 1929, many men of business suffered from what was near mass insanity, an unreasoning faith in the unending forward march of business, a reckless disregard of warning signals, a belief that I heard often expressed by men who are usually normal, sound and conservative, that we were in a new era in which the old tried and tested economic laws would not apply. Men were carried away by mounting, though financial prosperity. Judgments were warped. It was this frame of mind, leading as it did to undue factory and business expansion, to frenzied speculation in land and securities without regard to values and with a guileless belief that prices should be based on what values would be a decade hence,—it was this frame of mind, this unbelievable optimism, that was the cause of the reaction." Here, then, we have a "frame of mind" confidently indicated as "the cause" of a world-wide disaster and the consequent acute distress of many millions of men and women. Yet our official economists still keep up the pretence that their "science" needs no psychological foundation.

There is in America a lone worker who has much to teach the economists and who has even gained a hearing in some colleges as an occasional lecturer to classes in the departments of economics. Mr. Whiting Williams has spent many years working with his hands beside the artisan and the labourer, at the same time using his head to study them, their hopes, their fears, their aspirations ; in short, their motives. And he has embodied his observations and conclusions in several notable volumes.[1] This is the kind of study and the kind of knowledge which economics must make its inductive basis, rather than deduce its laws from a few misleading assumptions about a mythical economic man. How remote is the prospect that our universities will condescend to such studies ! Oxford is willing to show a sympathetic interest in the working man, to impart to him the rudiments of culture, to make him acquainted with the Republic of Plato and with the fact that philosophers cannot agree on the meaning of the word " truth ". But to study him systematically as a unit in the vast interplay of mental forces which is the economic and political life of the world, to do this would be to repudiate her traditions, especially the tradition that all we can hope to know of human nature is to be found in the writings of the philosophers of antiquity.

I submit that economics is not a science, but, rather, a chaos ; and that, in spite of the great amount of work done in the field, it

[1] Especially *Mainsprings of Men*, N.Y., 1923.

must remain a chaos until it can found itself on some systematic knowledge of these imponderables. And the same is true of all the so-called social sciences. As Mr. Loveday truly says : " It is well to remember when conditions are bad, men are without work, and children short of food, that the malady from which society is suffering is plain ignorance." But not ignorance of facts of the statistical kind, of which, as Sir J. Stamp tells us, we have multitudes, a plethora ; rather lack of such knowledge of the imponderables as will enable us to interpret these facts.

Dr. Othmar Spann, the philosophic historian of economic thought, concludes his survey of the field by saying[1] : " When we take a general survey of all the trends of economic science . . . we are led to the general conclusion that, from the historical standpoint, there is no unified body of economic doctrine. Nevertheless, the recognition that hitherto there has not existed any uniform system of economic doctrine must not lead us to doubt the possibility of a science of economics."

[1] *Types of Economic Theory*, 19th ed., Eng. Trans., London, 1930.

That is the conclusion on which I am insisting—that a science of economics remains an unrealized possibility.

Spann traces back the " classical " economics to Quesuay and his doctrine of a natural order. " This doctrine concerning the ' natural order ' is fundamental in two ways, inasmuch as therein the pursuit of self-interest is regarded as an economic postulate formulated in accordance with the theory of natural right, a system of economic individualism is for the first time established. Secondly, the persons who, in the economic regulation of their lives, act consistently because they are guided by the motive of self-interest, resemble atoms with fixed properties ; and the phenomena that result from their mutual contacts . . . are mechanically determined like those that result from the mutual contact of atoms. It follows that political economy, like the realm of material nature, is governed by purely mechanical laws." The individualist quasi-mechanical school thus founded has continued to claim the adhesion of most of the economists, including nearly all those of Britain and America. Spann severely criticises this school and sets over against it another type of economic thinking which he calls universalist, collectivist and romantic. The distinctive feature of this school is the recognition of society as an organized spiritual whole. Hegel and Fichte having revived the doctrine of a true community, " Adam Müller applied the newly acquired philosophical notion of community to economics, politics and sociology. In his view, the State was the aggregate of human affair, their interconnexion to form a whole ; it was something absolutely vitalized and spiritual,—it was a moral community." " For him the pre-requisite of all scientific thought concerning social questions was that it should penetrate every aspect of community life ; the economic, the political, the religious, and the moral .... Müller directs his gaze towards social life as a whole—in contradistinction to Adam Smith, who abstracts economics, to contemplate it in isolation ... Whereas Smith had regarded as productive that work only which was devoted to the making of concrete things, Adam Müller included under the head of production the 'ideally productive' achievements of the statesman and the artist. 'Ideal production'...the most splendid of a nation's assets, was according to Smith valueless when it was a question of estimating the total wealth of a nation. The words of the statesman, which would perhaps bring millions in actual money into existence ; the words of the priest or those of the artist, which might ennoble the heart or enlarge the imaginative faculty of the nation—these counted for nothing. I fancy I hear the objection . . . that you can only reckon up things that can be handled, measured, and weighed . . . I answer that 'the national existence in its widest possible scope is the true wealth of a nation'. Profoundly significant are his utterances concerning the fruitfulness of the labour of the State : 'An individual productive force can only produce in so far as it has itself been produced by a higher productive force. If the State ceases to produce, then, automatically, all the less productive forces cease to operate.' The 'products of all products' is the economic community, the natural economy."

But this way of conceiving economics

as but one partial aspect of the life of an organic whole has not been generally accepted by economists, and, in consequence, says Spann, " Economics, aware of its own poverty and perplexity, is degenerating into jurisprudence or even mere book-keeping, having little more to offer than useful descriptions of economic fact. . . . The main difficulty at the present time attaching to the study of economic theory is that it is so hard to transcend the narrow individualist Anglo-French doctrine which dominates the text books and economic literature in general." Again, he writes : " the very existence of a neo-liberal trend to-day (when all the Ricardian schools have proved so sterile in the field of theory) and still more the fact that this school should recently have become dominant, are manifest indications that our science is still talking the language of the eighteenth century."

It seems indisputable that the organic view of society or the nation as a spiritual or mental and moral unity, which is the distinctive feature of the universalist school of economics represented by Dr. Spann, does partly redeem that school and render it very superior to the school of classical and individualist economics. Yet I venture to think that Dr. Spann does not go nearly far enough in explicit recognition of the need for an adequate psychological basis of economics. He writes : " Unsound, finally, is the quantitative and mechanical method which was the outcome of Smith's and Ricardo's general attitude. For if the essence of economic activity consists not in quantity of labour, or quantities of goods, but in the purposive articulation of functions in the extant aggregate of the economic body (or the functioning economic organism) then there is no more place in economics for laws of mechanical causation than there is place for mechanical causation to intrude among the laws of logical thought (which are likewise purposive and not mechanical). Economics has, indeed, an inner and unambiguous determinism ; but that determinism is not mechanically causal. The ' laws ' of supply and demand therefore, like the ' laws ' of prices, are neither natural nor primary laws, being solely the expression of purposive, articulate, functional interrelations." In this and similar passages he betrays the inadequacy of his own system. He is right in insisting on the organic unity of society within which economic activity proceeds, and so far points to one of the grand defects of the classical economics. But in declaiming against the mechanical causation of that school he makes a false diagnosis. Self-interest, made by that school the sole and and universal economic motive, is not a mechanical principle : it is thoroughly purposive. But it was combined by the classical economists with the false principle of hedonism and with the quasi-mechanical psychology of the association school of the Mills, Bain and Spencer. This provided the minimum indispensable psychological basis of the classical economics ; but it was a psychology of a spurious and very misleading simplicity. What is needed is an economics that shall frankly recognize the rich complexity of the motivation of human action, of economic no less than of political and of all other forms of social activity : shall recognize that " the forces and tendencies " of which economists write so obscurely are in the last analysis mental forces, desires, strivings, seekings, purposively and more or less intelligently directed thinking, all springing from the hidden " main springs of men ", their loves and hates, their personal and group loyalties and family affections, their patriotism, their fears, their tastes, their ambitions, their ideals, their moral and religious and æsthetic aspirations, their compassions, their sentiments of honour and justice and humanity, their passion for liberty and equality and human brotherhood, no less than from lust and " herd instinct " and sheer greed or " self-interest ".

Psychology, even in its present rudimentary and chaotic condition, is capable of rendering great services to the social sciences, as several notable books by Professor Graham Wallas have shown. One main root of the present trouble is that psychology cannot now, and perhaps never will be able to, present conclusions in the form of simple formulae that may be applied to social problems by any man. The social application of psychological truths will always require to be made by men trained in and habituated to thinking in psychological terms. At the present time a belief in the value of psychology is widely diffused in America. This, I believe, has done much in the last decade to bring about much improved relations between capital and labour and has thus contributed not a little to the economic prosperity envied by all the world up to October, 1929. (The campaign of high-powered salesmanship was only one, and one of the most questionable, instances of such applications ; of questionable ultimate value, although it did undoubtedly do much to promote the boom.) But it has led to many crude and altogether undesirable and unsuccessful applications. The most notable, perhaps, of these was made by the engineer-president. He seems to have laid to heart the formula that confidence may be produced or promoted by suggestion and that, for effective suggestion of this kind, no more is necessary than reiteration of an optimistic assertion by persons in authority. When the slump of the stock-market began, the President and those near to him at once gave out to the public reassuring predictions, and the market promptly fell again. This process was repeated several times, without any other striking effect beyond a great reduction of the public confidence in the wisdom of the administration. The moral of this unfortunate incident is not that psychological factors may be safely ignored ; rather it illustrates their overwhelming importance and the fact that they require very delicate expert handling ; that no crude rule-of-thumb methods will suffice.

It is the same story with all the so-called social sciences. They are backward, rudimentary, not only because they are starved and neglected, but also and chiefly because they have not for their foundations any science of the imponderables, those human energies with which they are almost exclusively concerned ; and, because the exponents of these sciences, so far as a large proportion of them are concerned, are not even aware of this need, this lack of an indispensable foundation.

Consider for one moment so-called political science, or the science of government. Is it not almost as chaotic as economics, as incapable of prediction, of explanation and of trustworthy precepts ? And is not the ground of its defects the same ? When it aspires to predict, it has nothing more than the bald statement that history repeats itself—or does not repeat itself—according to the fancy of the author. When it seeks to explain, it invents *ad hoc* various strange instincts.[1] When it is a question of prescribing, it can offer only a few vague and abstract generalities.

I take up the first book on political science that lies to hand. It is *Sinon, or the Future of Politics*, by Mr. A. E. Mowrer. It begins with a citation from Lord Bryce's *Modern Democracies*. "The tendencies of human nature are the permanent basis of the study which gives to the subject called Political Science whatever scientific quality it possesses." Mr.

---

[1] This is an old trick of writers on political topics who aspire to be philosophical; and it is still the prevalent fashion. In the recent highly and rightly praised work on England by Dr. N. Dibelius, I find that in one chapter dealing with English characteristics he uses as explanatory principles the following: The conservative, the monarchical, the gentlemanly, and the religious instincts, the business and the sporting instincts, the instinct for power, the irrational instinct to occultism, the instinct for the non-rational, as well as innate mistrust, and instincts of the beast of prey. We shall make no progress in political science and in the understanding of racial and national characteristics until this fashion of too easy explanation by the *ad hoc* postulation of special instincts shall give place to more resolute grappling with the psychological problems which confront on every hand the writer on these topics.

Mowrer, having put this text, this sound prescription, at the head of his discourse, opens it with the following sentence. "All men are born politicians ; that is, born into a relentlessly political society, and equipped with a political instinct that only in rare individual cases is repressed or surpassed. Moreover, the political instinct is not only ubiquitous to historical mankind, but is apparently shared with other socially minded creatures." He makes no attempt to define this alleged and, if real, all important " political instinct ". Indeed, he makes it obvious that he has never considered what he means by " instinct " and still less what he means by "a political instinct ". Instead, he proceeds to expound the whole science of politics in fourteen small pages. Here he follows and expounds Umano, who, he tells us, " is the profoundest student of politics in modern times." And here is the essence of Umano's teaching : " Political study can claim attention as a branch of science only if it build upon elements common to all recorded experiences of a political type. In point of fact, every political situation does contain two constants stout enough to bear the weight of general conclusions. These constants are (1) an individual and collective instinct for dominion over others. [Is this the alleged ' political instinct ? ' Our author does not trouble to tell us.] (2) The means or forces whereby such dominion can be sought, consisting of (*a*) a force of mind ; (*b*) a force of bodies, and (*c*) a force of wealth or property. Politics, *therefore* [note the deduction process], is simply the struggle resulting from the efforts of each human being to induce, persuade, or constrain his fellows to do his will, and conversely, not to be induced, persuaded or constrained into doing theirs. Each person, in addition to a will to power, possesses from childhood until death a force of mind with which to think and persuade ; a force of body with which to act, to attract and to constrain the bodies of others ; a force of wealth with which to extend, aid and increase the other forces ... These forces are inherent in the world, and there are no others conceivable." It is as simple as that ! So much for the " tendencies of human nature ", the study of which, Lord Bryce declared, gives to political science whatever scientific quality it possesses. They are two and two only, " a will to power " and " a force of mind ", and there is nothing more to be said. No wonder that our author quickly arrives, by deduction from these premises, at some startling conclusions (following always the profoundest political thinker of modern times) ; such as that might makes law ; " justice is rigorous enforcement of existing law, injustice the lack of enforcement or lax enforcement " ; and, " The much-sought causes for the downfall of States can be only two : insurrection from within as the result of despotic government ; conquest from without as the inevitable outcome of international anarchy."

The prevailing confusion in political thinking is nowhere more deplorable than in relation to nationalism and internationalism. Almost without exception these are treated as mutually exclusive alternatives. Even Mr. Norman Thomas, generally so clear-headed, does not escape this fallacy, as when he writes : " the desirable alternative to nationalism is internationalism, not imperialism." [1] Surely, it is obvious that

[1] *America's Way Out*, N.Y., 1931.

internationalism means, or should mean, not the destruction of nations, but closer co-operation between nations for the common good of all : the development of international law and its sanctions, which in turn means some abatement of the claims of nations to absolute sovereignty and something like a federation of nations or states. It should be obvious to the meanest intellect that to approach the task of creating a new political organization of the world by way of first destroying such organization as we have, organization mainly in national groups, would be to render the attempt perfectly hopeless. If such organization could be set up, the first great task of the world-government would be to organize regional groups to assume the place and functions of the destroyed nations. World government by a single centralized bureaucracy is, surely, a crazy ideal !

Here then is the only road to remedy the parlous and ever more dangerous state of our civilization. We must actively develop our social sciences into real sciences ; and, in order to do that, we must first create a science of the imponderables ; in short, of human nature and its activities. But perhaps that is impossible. The school of thought pre-dominant in this country asserts the impossi-bility. They tell us, by implication, if not in so many words, that all we can ever hope to learn of the imponderables is to be found in the works of Plato and Aristotle. This seems to me a doctrine of despair. For we have tried the intensive study of these great authors for many generations ; and it has not met our needs. Such study is admirable as a method of attach-ing our civilization to its foundations ; but of little avail to correct its lop-sidedness, to furnish a basis for the social sciences we need. To despair of achieving systematic knowledge of the imponderables before we have made a serious and sustained effort would be to manifest a craven spirit. What, then, in practical terms, is the remedy. I can give my answer most concisely by suggesting what I would do if I were Dictator. I would, perhaps, permit the continuance of physical research by the industrial corporations. But I would by every means seek to divert all our most powerful intellects from the physical sciences into research in the biological, the human and the social sciences ; and our universities should be the main seats of such research.

Dr. Abram Flexner, in his recent book on Universities, writes : " Somewhere, away from the hurly-burly of practical responsibility and action, the social and political problems involved in these disorders must be exposed. The 'great society' must, and wants, to understand itself—partly as a matter of sheer curiosity, partly because human beings are in a muddle and cannot get out unless they know more than they now know. Towards fundamental knowing the newspapers cannot help much ; men of action, politicians, and business men help but slightly. They themselves know so little, they are not disinterestedly concerned with finding out ; they have usually their own axes to grind. Almost the only available agency is the University. The Universities must shelter and develop thinkers, experimenters, inventors, teachers, and students, who, without responsibility for action, will explore the phenomena of social life and endeavour to understand them . . . No university in any country has made really adequate provision or offered really adequate opportunity and encouragement." . . . " The task, then, of finding a basis and providing a methodology for the social sciences is to-day more pressing than it has ever been because of the accelerated rate of social change and relatively more rapid progress in the physical and biological sciences . . . If the University does not accept this challenge, what other institution can or will ? "

It is obvious that much might be done merely by reallocation of the financial resources available for the support of research. But new funds ear-marked for the social sciences are needed. Yet the most liberal allocation of funds will not in itself suffice. We need a new scale of values in the public mind, a scale in which such prestige shall attach to the social sciences as hitherto physical science alone has enjoyed. We need many workers, but, more urgently still, we need intellects of the first order, devoted to the study of man and society as they are now devoted to the study of stars and atoms. And we need to avoid premature applications to practical problems. The urgent need and desire for practical guidance from the psychological and social sciences has been their bane and their undoing from their first inception. Locke, Hobbs, Bentham and all the school of utilitarians, in modern times illustrate my point vividly. They were primarily social philosophers ; that is to say men concerned to devise and advise social and political improvements or to justify existing institutions. In their haste towards these ends they neglected the prior task of laying truly the scientific basis upon which alone such philosophic structures can be safely built. Our civilization has paid for the neglect in a century of discussion of " social contracts " and " economic men " and other myths, while social development has gone its zig-zag chaotic course, leading to our present parlous state. The greater urgency of men's interest in what should be, rather than in systematic knowledge and analysis of things as they actually are, has maintained the subordination of the sciences of man and of society to philosophy. In our academies these sciences have in the main been left, until very recently, in the domain of the philosophers ; and in their hands the scientific problems have received scant attention. Thus all the science of conduct, the problem of human motivation, was left to the ethical philosophers, whose prime interest was in the question how men should conduct their lives in general and in face of various situations ; and the fact that acceptable answers to all such questions can be found only in the light of scientific knowledge of man, especially of the principles of motivation, was almost ignored. The same subordination of the scientific to the philosophical problems has been even more marked and injurious in the sphere of economics and politics. A further evil has been the divorce of both economics and politics from ethics. Abstraction is necessary in the sciences ; but is apt to be very misleading and, in the philosophical discussion of rights and duties, disastrous. The classical instance of the procedure I am deprecating is Jeremy Bentham's. He opens his famous treatise on morals and politics with a sweeping assertion about human motivation (the hedonist assertion, now generally recognized to be false) and proceeds to build up his doctrine deductively from these false premises, thus setting in train all the long course of errors of the utilitarians and the Manchester school.

We need to make a fresh start in all these disciplines with a full and frank recognition of their need for a psychological foundation ; and, as Dr. Flexner insists, it is to the Universities we must look for this development. The London School of Economics seems to express recognition of this need ; but in that institution research needs far greater emphasis. The Yale Institute of Human Relations is another such expression. I do not pretend to stand alone in making this demand. I seek here merely to state the need, now beginning to be widely recognized, emphatically and without the reserve which regard for one's colleagues so strongly calls for. If I have seemed to reflect upon any of them, that is a misreading of my essay. The present state of affairs is the outcome of an unfortunate course of history for which no individual and no class is to blame. But it is one which before all things calls for clear thinking and plain speaking.

Especially, I would at first concentrate the attack upon Anthropology, the science of man, conceived in the broadest way. It would take some twenty years to train the personnel and get them working on their problems. Then, as they began to bring in results, part of the available intellectual energy would be concentrated in the effort to build the social sciences, especially a science of economics, on the basis provided by the anthropological research. On a hopeful view, another twenty years would elapse before substantial progress along this line might be expected. That would bring us to the year 1970, or thereabouts.

Can we afford to wait so long ? Can our civilization survive in the meantime ? I don't feel sure ; but I hope it may : for I can see no alternative measures that offer hope of its salvation.